Health 59

秸秆的妙用
Jobs and Beauty from Straw

Gunter Pauli

冈特·鲍利 著

李欢欢 译

丛书编委会

主　任：贾　峰
副主任：何家振　郑立明
委　员：牛玲娟　李原原　李曙东　吴建民　彭　勇
　　　　冯　缨　靳增江

丛书出版委员会

主　任：段学俭
副主任：匡志强　张　蓉
成　员：叶　刚　李晓梅　魏　来　徐雅清　田振军
　　　　蔡雩奇

特别感谢以下热心人士对译稿润色工作的支持：

姜竹青　韩　笑　杨　爽　周依奇　于　哲　阳平坚
李雪红　汪　楠　单　威　查振旺　李海红　姚爱静
朱　国　彭　江　于洪英　隋淑光　严　岷

目录

秸秆的妙用	4
你知道吗？	22
想一想	26
自己动手！	27
学科知识	28
情感智慧	29
艺术	29
思维拓展	30
动手能力	30
故事灵感来自	31

Contents

Jobs and Beauty from Straw	4
Did you know?	22
Think about it	26
Do it yourself!	27
Academic Knowledge	28
Emotional Intelligence	29
The Arts	29
Systems: Making the Connections	30
Capacity to Implement	30
This fable is inspired by	31

水稻种子拼命地在找一个地方来处理秸秆，那些秸秆是由废弃的水稻茎秆组成的。胡萝卜一直在陪着他。

水稻种子叹了口气，说道："你知道的，以前有很多房子是用秸秆盖的，但近来，人们更喜欢用钢筋、水泥和红砖盖房子了。"

A rice seed is desperately looking for a place to get rid of the straw made of leftover rice stalks. A carrot is accompanying him on his search.

The rice sighs. "You know, there used to be so many houses built with straw, but lately people prefer steel, concrete and red bricks."

……找一个地方来处理秸秆……

... looking for a place to get rid of the straw ...

稻草屋

Straw-bale houses

"是呀，"胡萝卜说，"不过有很多人对稻草屋感兴趣。"

"当然了，那很好，但是每年秸秆产量有几百万吨，却没有造上百万座稻草屋。连日本的榻榻米都越来越不受欢迎了。"

"Yes," comments the carrot, "but there are lots of people interested in straw-bale houses."

"Of course, and that's great, but we produce millions of tons of straw every year, and there are not millions of straw-bale houses. Even Japanese tatami mats are losing favour."

"你说得对。但起码中国人在利用秸秆种植蘑菇呀。"

"他们确实有变废为宝的传统。要不然他们怎么能养活十几亿人口呢？"

"You're right. But at least the Chinese use straw to farm mushrooms."

"They do have a culture of converting waste into food. How else would they feed more than a billion citizens?"

用秸秆种植蘑菇

straw to farm mushrooms

……短杆水稻？

… rice with shorter stalks?

"水稻,你听说了吗?在埃及首都开罗,人类正试图改变水稻的基因,生产短杆水稻。"

"为什么要这么做呢?"

"Rice, have you heard that in Cairo, the capital of Egypt, the humans are fiddling with your genes to make rice with shorter stalks?"

"Why would you ever want to do that?"

"呃，人类曾经用秸秆造房子，但现在这不流行了。因此他们决定焚烧堆积如山的秸秆。"

"焚烧？！人类什么时候才会彻底认识到焚烧并不是解决之道呀？"

"Well, they used straw for construction, but it wasn't popular. They had so much piled up that they decided to burn it."

"Burn it?! When will humans finally learn that burning is not a solution?"

决定焚烧秸秆

Decided to burn it

影响孩子们的健康

Risks the health of children

胡萝卜明智地点点头。"这么大量的焚烧,空气里弥漫着浓烟,没人能看清尼罗河的对岸。"

"别说了。焚烧秸秆会影响孩子们的健康。"

The carrot nods sagely. "The burning was so intense, the air got thick with smoke and no one could see across the Nile."

"Don't tell me. Burning straw risks the health of children."

"绝对不利,越来越多的孩子得病,呼吸困难。人类可以想出的唯一解决办法就是改变水稻品种,只种植短杆水稻。"

"所以,人类又想干涉我,就像以前他们给我加入胡萝卜基因一样吗?我有一个更好的主意。埃及人和意大利人聊过吗?"

"为什么他们要这么做?"

"Absolutely, more and more children got sick and couldn't breathe. The only solution they could come up with was to modify rice and only grow the type with short stalks."

"So, these humans want to interfere with me like they did when they added some carrot genes to me? I have a better idea. Have the Egyptians talked to the Italians?"

"Why would they?"

埃及人和意大利人聊过吗?

Have the Egyptians talked to the Italians?

利用秸秆来制造塑料

Make straw into plastics

"意大利人种植水稻有2000多年的历史了。他们生产粮食的能力巩固了罗马帝国的战斗力。"

"埃及人并不想打仗——他们为什么要和意大利人聊聊呢?"

"噢,意大利人已经学会了如何利用秸秆来制造塑料。"

"The Italians have been farming rice for more than 2 000 years. Their capacity to produce food powered the war machine of the Roman Empire."

"The Egyptians aren't thinking about war - so why talk to Italians?"

"Well, the Italians learnt how to make straw into plastics."

"抱歉,水稻,不过塑料是由石油制成的。"

"过去确实如此,但现在人类可以把秸秆分裂成微粒,就像炼油厂提炼石油一样。"

"如果那是真的,那就让大家多吃大米,制造更多的塑料吧!我想起来了,我知道有人甚至可以用水稻做出珠宝……"

……这仅仅是开始!……

"Sorry, Rice, but plastics are made from petroleum."

"They used to be, but now humans can crack rice straw into dozens of molecules, just like a refinery does with oil!"

"If that's possible, let's all eat more rice and make plastic! Now that I think about it, I know someone who can even make jewellery out of rice …"

... AND IT HAS ONLY JUST BEGUN!...

……这仅仅是开始！……

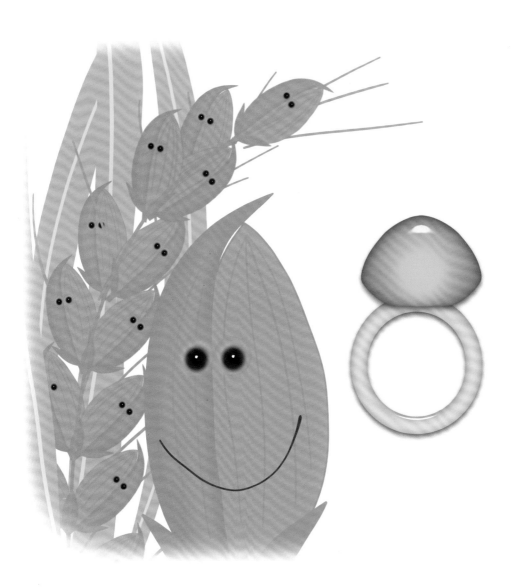

...AND IT HAS ONLY JUST BEGUN!...

Did You Know?
你知道吗？

Straw-bale houses are built to last. There are straw-bale houses that are 140 years old, and those constructed today could last for at least 100 years.

稻草屋经久耐用。有些稻草屋已有140年的历史。今天建造的稻草屋至少能使用100年。

A single rice straw is not strong, but baled rice straw can last for decades, provided that it is protected from rain and humidity.

一根秸秆不结实，但成捆的秸秆只要不受雨淋，不受潮，就可以用上几十年。

Tatami flooring was originally reserved as seating area for the most important aristocrats in Japan. It was popularised in the 17th century, and only lost favour after the Second World War, though most Japanese homes still have one tatami room.

最初，榻榻米是为日本最负名望的贵族预留的专属座区。榻榻米盛行于17世纪，二战之后才渐渐不受欢迎，不过大多数日本家庭依旧保留着一间榻榻米屋。

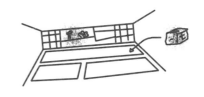

Straw mushrooms are also called paddy straw mushrooms and have been farmed in China for hundreds years. These mushrooms could secure more food than could be supplied by rice.

草菇也叫杆菇，已在中国种植了几百年。这些菇类所能提供的食物比大米提供的还多。

The first book describing 11 edible mushrooms was written in 1245, and described mushrooms in Chekiang Province (now Zhejiang Province), China. The first time straw mushrooms were documented was in 1822.

第一本记载了11种可食用菇的书写于1245年，描述的是中国浙江省的菌菇。1822年，第一次出现草菇的文字记载。

Every year there is as much straw produced by plants as there is petroleum extracted from the Earth.

各种作物每年产出大量的秸秆，几乎相当于每年开采的石油量。

Straw is rich in cellulose, which can be converted into plastics. We could potentially make all the plastic we need from the waste that is burnt today.

秸秆富含的纤维素可以转变成塑料。我们今天焚烧的废弃物很有可能生产出我们需要的所有塑料。

The company Novamont in Italy pioneered the methods of converting agricultural waste into the chemical products that they and society need.

意大利诺瓦索特公司将农业废弃物制成人类和社会所需的化工产品，他们在这一领域处于领先地位。

Think About It

Would you burn straw if you knew that you could make houses, plastics, food and fuel from it?

如果你知道可以利用秸秆建造房子、生产塑料、种植食物、提供燃料，你还会焚烧秸秆吗？

你更喜欢科学家做哪件事：是改变水稻基因，发明短杆水稻；还是充分利用秸秆的所有可能用途？

What would you prefer that scientists did: genetically modify rice to have short stalks, or to capitalise on all the possible uses for naturally occurring straw?

Would you like to live in a straw-bale house?

你愿意住在稻草屋里吗？

焚烧废弃物是个好办法吗？还是应该利用我们所认为的废弃物，努力创造产品和工作机会？

Is burning waste a good solution? Or should we try to create products and jobs with what we consider to be waste?

Do It Yourself!
自己动手!

You may not have rice straw lying around at home, but you'll certainly have drinking straws handy. After plastic bags, drinking straws present one of the greatest problems for aquatic life. We need to find something to do with them! There are thousands of ideas, but try to think of your own idea for straws that you can do at home. My favourite is making a straw aeroplane. Figure out how to do it yourself – it flies better than a paper plane!

你家附近可能没有秸秆，但吸管肯定随手可得。水生生物面临的最大问题之一就是吸管，仅次于塑料袋。我们必须找到处理它们的方法！虽然已经有成千上万个方法，但请努力想出自己的方法，你在家就可以实行的方法。我最喜欢的方法是做成吸管飞机。想想看，你该怎么做——它飞得比纸飞机棒多了！

TEACHER AND PARENT GUIDE

学科知识
Academic Knowledge

生物学	草菇菌盖被顶破之前的卵形期是最佳采收期，此时的草菇含有最丰富的氨基酸；空气湿度超过20%时秸秆分解；秸秆适合用于建造牛舍，不适合用于喂养；除了甘蔗和玉米，大米是最广泛食用的主食。
化学	成熟的水稻作物40%的氮、30%的磷、80%的钾和40%的硫存在于秸秆中；秸秆富含锌元素，为水稻的硅元素平衡发挥重要作用；燃烧秸秆导致所有的氮和50%的硫流失；水稻外壳的硅和碳可以被转化成金刚砂。
物理	一根秸秆的抗压强度很弱，但成捆的秸秆可以和任何建筑材料媲美；成捆秸秆是天然的阻燃剂；谷粒通过水稻外壳的小孔进行呼吸，这意味着外壳的硅元素可以通过气孔渗透出来，应用于工业。
工程学	草砖房利用小麦、水稻、黑麦和燕麦的茎秆作为结构要素或隔热材料，或既是结构要素又是隔热材料；传统的埃及建筑使用尼罗河的泥和秸秆混合建造而成。
经济学	草菇种植量全球排名第三，仅次于蘑菇和香菇；秸秆可以转化成生物碳；水稻秸秆水解后可转化成生物柴油；木糖喂养的真菌可生产乙醇；埃及水稻农场平均每公顷产量9.5吨，而世界水稻平均产量仅有4.5吨。
伦理学	秸秆能够用来种植蘑菇、生产塑料、生产清洁燃料和补充土壤，这种情况下，如何能够证明焚烧秸秆或者改变基因生产短杆作物是合理的？
历史	非洲自旧石器时代起就建造草房；马里共和国的杰里大清真寺是世界上最庞大的泥砖和秸秆结构建筑；公元前7000年至公元前3300年期间，南非已使用泥砖；8200至13500年前，中国珠江流域已经移植水稻；非洲水稻品种于3500年前移植于尼罗河流域；水稻经由西亚传入欧洲和美洲。
地理	草菇种植于湿热的东南亚地区；泥砖在西班牙叫做风干土坯，广泛用于西班牙和拉丁美洲。
数学	25亿吨秸秆可以转化成：(1) 多少蘑菇？(2) 多少吨塑料？(3) 多少升乙醇？这些行业可以创造多少个就业机会？
生活方式	在电影《幻想曲》里沃特·迪斯尼给草菇戴上棕色草帽，配上了舞动的双脚，成了快乐蘑菇；蘑菇已经逐渐成为世界各地主食的一部分。
社会学	现代科学对转基因感兴趣，然而关于废弃物管理的研究同样应被认为科学；在亚洲，水稻被视为好运和繁荣的象征。
心理学	如果总是抽到下签，你会觉得自己运气一直不好。
系统论	秸秆是农业废弃物，与其烧掉，还不如把它转化成有用的产品，正如几百年来一直做的。

教师与家长指南

情感智慧
Emotional Intelligence

水稻种子

水稻种子分享了她的担忧。她充分认识到除了大米，水稻产出了更多的东西：大量的秸秆和谷壳。她察觉到她的副产品越来越不受欢迎。水稻种子认为种植蘑菇是一个必不可少的选择。她非常诧异人类为什么要发明短杆水稻，当得知通过焚烧来处理过量的秸秆时，她感到紧张。她对焚烧的后果很敏感，尤其是对孩子们健康的影响。她觉得有责任积极寻找解决方案。她知道一个现成的解决方法，并坚持用这种方法把秸秆转化成一系列天然化学品。水稻种子的建设性提议激起了胡萝卜更多的积极思考。

胡萝卜

胡萝卜有着积极乐观的态度，想要鼓励水稻种子。当水稻种子对他的积极言语不做回应时，他提供了更多的积极消息。他俩试图寻找实用的可持续性解决方法。首先他们否定了人类的解决方法，因为这些解决方法引起了更多的新问题：焚烧不可取，事实说明焚烧与人类健康风险相关；转基因不适合，会导致资源永久流失。胡萝卜没有跟上水稻种子的思路，一时无法理解从石油到生物塑料的跳跃。然而，一旦他明白了，立刻恢复了积极的心态，提醒说水稻甚至可以制成珠宝首饰。

艺术
The Arts

现在我们要加大难度了。准备好了吗？我们要开始在大米上作画了。大米很小，因此你得有一枝很细的铅笔。试着在大米上画一条线，手指千万不要抹掉画好的线。感到受挫之前，再试一次吧。大米作画艺术非常盛行。如果一直画不好，你可以玩另一个画画游戏！取100粒大米，看看它们的色泽有什么不同。把它们放在太阳底下，你会看到细微的差别。把米粒分类，设计一个由明到暗的圆圈。

TEACHER AND PARENT GUIDE

思维拓展
Systems: Making the Connections

全球每年产生的秸秆总量相当于全球石油开采量。秸秆有很多用途，首先是种植蘑菇，这已有200年左右的历史，促进了中国的食品保障。如果所有秸秆都用于种植平菇或草菇，我们只需花一半的力气，就能多生产12.5亿吨食物。这意味着我们可以把2020年的粮食预测产量从21亿吨提高至33.5亿吨，让我们能够消除饥饿。秸秆不易消化，却可以在酶的作用下分解成富含氨基酸的真菌，成为极好的动物饲料。不仅如此，秸秆还可以用于塑料制造。如今，秸秆经过生物炼制——类似于把石油分子转换成酯，可以制成具备各种价值和功能的商品。秸秆还可以转换成燃料、乙醇或者生物柴油；由此产生的废弃物又可以被视为另一种资源。水稻秸秆用于建造房屋已有上千年历史，秸秆捆在耐用性、防火性和能源效率方面受到认可。谷壳富含二氧化硅，由此提取的硅元素能转化成各种产品，包括电池的硅电极和珠宝首饰。目前受到推崇并广泛应用的秸秆处理方法不仅是不可持续的，还剥夺了人类创造性地利用原材料的可能性，秸秆利用是满足食物、住房、能源等工业和社会需要的重要途径。这就把改变水稻基因，生产短杆水稻的决策放在质疑的聚光灯下。社会和决策者是否能容忍，仅因某家公司的利益就丢弃淘汰如此常见的资源——一种能创造产品、价值和无数就业机会的资源？

动手能力
Capacity to Implement

把所有能用秸秆制成的产品列成清单。这则寓言已提供了一些可行的应用，可一旦你想到了塑料，你还会大有收获。现在，罗列出哪些不可持续性产品可用水稻产品代替，或者应该创造哪些新产品，以致地球资源不被耗尽。我们建议，企业要更善于利用地球现有的资源。

教师与家长指南

故事灵感来自

伊江玲美
Remi Ie

伊江玲美认为自己是一名可持续发展型的企业家。她热衷于改善农民生活水平，提高农产品行业的竞争力。她与冲绳科技研究所合作，尝试了多项创新，包括创立美人鱼咖啡馆，那儿的有机食品由水稻、小麦和糖蜜制成。伊江对谷壳很好奇，从商人那儿学会了怎么把二氧化硅转变净化成珠宝首饰。这些首饰只是用水稻废弃物制成的，却透着珍珠的光泽。

更多资讯

http://www.ecobuildnetwork.org/projects/research/straw-bale-test-program/straw-bale-fire-test-video

http://www.sciencedirect.com/science/article/pii/S0008443397000177

http://www.nec.com/en/global/eco/featured/bioplastics2/contents2.html

http://theverybesttop10.com/things-to-do-with-drinking-straws/

http://www.newscientist.com/article/dn23839-ricehusks-could-make-much-longerlasting-batteries.html#.VKfXlKCHqwA

图书在版编目（CIP）数据

秸秆的妙用：汉英对照／（比）鲍利著；李欢欢译．－－上海：学林出版社，2015.6
（冈特生态童书．第2辑）
ISBN 978-7-5486-0851-6

Ⅰ．①秸…　Ⅱ．①鲍…　②李…　Ⅲ．①生态环境－环境保护－儿童读物－汉、英　Ⅳ．①X171.1-49

中国版本图书馆CIP数据核字（2015）第086067号

————————————————————————————

© 2015 Gunter Pauli
著作权合同登记号 图字09-2015-446号

冈特生态童书
秸秆的妙用

作　　者——	冈特·鲍利	
译　　者——	李欢欢	
策　　划——	匡志强	
责任编辑——	李晓梅	
装帧设计——	魏　来	
出　　版——	上海世纪出版股份有限公司 学林出版社	
	地　址——上海钦州南路81号　电话／传真：021-64515005	
	网　址——www.xuelinpress.com	
发　　行——	上海世纪出版股份有限公司发行中心	
	（上海福建中路193号　网址：www.ewen.co）	
印　　刷——	上海图宇印刷有限公司	
开　　本——	710×1020　1/16	
印　　张——	2	
字　　数——	5万	
版　　次——	2015年6月第1版	
	2015年6月第1次印刷	
书　　号——	ISBN 978-7-5486-0851-6/G·300	
定　　价——	10.00元	

（如发生印刷、装订质量问题，读者可向工厂调换）